NISTIR 7538

USER'S GUIDE FOR THE QUALITY OF DESIGN TESTING TOOL AND THE CONTENT CHECKER

KC Morris
Simon Frechette
Puja Goyal
Josh Lubell

Boonserm Kulvatunyou
Salifou Sidi Malick
Nicolas Brayard
Severin Tixier

Manufacturing Systems Integration Division
Maufacturing Engineering Laboratory

October 2008

U.S. DEPARTMENT OF COMMERCE

Carlos M. Gutierrez, Secretary

National Institute of Standards and Technology

Patrick Gallagher, Deputy Director

Contents

1. Abstract 1
2. Introduction 1
3. Test Procedure: Quick User's Guide 3
   3.1 Test Requirement Management 3
   3.2 Test Case Management 4
   3.3 Test Profile Management 5
   3.4 NDR Profile Management (QOD only) 6
4. Test Materials 8
   4.1 Test Document 8
   4.2 Test Requirement 8
   4.3 Test Case 13
      4.3.1 Schematron Considerations 14
      4.3.2 Java-based Expert System Shell (JESS) Considerations 17
   4.4 Test Profile 21
   4.5 Test Result 21
5. Conclusion 23
6. References 23
7. Acknowlegements 25
8. Disclaimer 26

1 Abstract

This document describes the operation and usage of the Quality of Design and the Content Checker Testing Tools. These tools were developed at the National Institute of Standards and Technology (NIST) to support people in developing standards for the exchange of data based on the Extensible Markup Language (XML) and the related standard XML Schema from the World Wide Web Consortium (W3C)[1]. The tools are web-based and provide a collaborative environment for users to develop tests of their data exchange specifications and data based on those specifications. Additionally, tests for some 3rd party guidelines for XML Schema are available through the web-based systems. This document is a user's guide for people using those systems either to develop their own tests, to use the tests that are available, or to augment the existing tests with their own.

2 Introduction

Software is developed to solve specific tasks that are time intensive for human beings but that are well defined so that they can be consistently and repetitively performed. In other words, software typically addresses a specific problem. However, as soon as a task is automated then the tasks surrounding it become ripe for automation as well. This joining of one piece of software with another is called system integration. Often that integration becomes so transparent that the features of the two pieces of software become one application; other times the integration remains less transparent depending on the circumstance. A common approach to integration is data exchange. In other words, the data from one system is given to another system, which is able to perform a different set of functions on that data and/or may reside in a different location under different ownership. A common example of this is in collaborative document editing: a document may be developed by one person using a software application and the data or the "document" can then be given to another person to add to or change using a similar but different piece of software.

The XML family of standards [2] has been developed to address the need for a format in which to transfer data between software systems. Many groups have adopted these standards as a basis for defining their own formats for data exchange. These specifications define the terms to be used to exchange data between different systems; however, the definition of these specifications is a challenge in itself. The more diverse the systems sharing the data are, the more complex the task is of defining the specification for data exchange. In order to facilitate the specification of the data, groups provide guidelines on how the XML standards should be used in their contexts. Many such guidelines have been published and more are unpublished. These guidelines are generically referred to as XML Schema Naming and Design Rules or NDRs. In this paper we use the term XML Schema to refer to the XML Schema Specification from the World Wide Web Consortium [3]. Similarly, once specifications have been defined for data exchange, how the data from a given system will be expressed in that specification is also subject to guidance. This guidance is less often published in formal documentation but is nonetheless important for accurate data exchange. For a more complete discussion of the data exchange process for systems integration see [4, 5].

NIST has developed two tools to aid in this process: the Quality of Design (QOD) Testing Tool and the Content Checker. The QOD tool aids in the definition of tests for XML Schema specifications [6]. The Content Checker tool aids in the definition of tests for the XML instance data. This document describes these tools from a user perspective. The illustrations for the document are the user interfaces for the tools. The document consists of two parts: the test procedure and the test materials. The user interfaces, characterized by the test procedure and test materials, for the two tools are very similar and the text describes where they diverge.

Both tools are web-based systems and are available at http://www.mel.nist.gov/msid/XML_testbed/index.html [7]. They may be accessed in two ways: 1) As an guest user you can check your files against tests defined in the system. See Testing as a Guest User. 2) You may also request an account to use more of the features. With an account you may

- submit your own rules and tests to use in checking schemas,
- share those tests with other people (including anonymous users),
- test XML schemas, and
- track the results from testing your schemas.

This document is intended for users with an account and describes the full range of features available in the system. Guest users may also find it helpful but will not have access to all the features.

When logged into an account, the menu bar pictured below is shown on the left of the screen. It includes all the functions described below. When accessing the system as a guest user an abbreviated menu is shown.

About QOD
Welcome
Test Procedure
Testing as Guest
Example Page
My Account
Login
Change Password
Previous Test Results
My Test Results
Test Requirement
List TRs
Register TR
Test Case
Schematron
JESS
List TCs
Add TC
Test Profile
List TPs
Create TP
NDR Profile
Import
Export

Note that items underlined in this document are available as hyperlinks in the online documentation.

3 Test Procedure: Quick User's Guide

This section describes how to use the XML Schema Quality of Design (QOD) and the Content Checker Tools. Please see the Test Materials section to learn more about the materials in the system: Test Requirement, Test Case, Test Profile, and Test Result. This section describes the management of these items through the functions in the menus listed on the left-hand side of the screen.

In brief, a Test Requirement is a statement of a requirement on an XML schema (in QOD) or a XML document (in the Content Checker). A Test Requirement is verified by one or more Test Cases which are computationally encoded versions of the requirement. Test Profiles are a collection of Test Cases that may be executed in bulk on an XML schema or document. When a file is "tested" with a Test Profile, all the Test Cases in that profile are executed against the file and the results are reported in a browser window and, if the user has an account, stored in the system for future reference.

In order to use all these features you will need an account on the system. If you do not have an account, you can still use the system to test your schemas with existing rules but you will not be able to create and manage new tests. Go to Testing as Guest to use the existing rules.

3.1 Test Requirement Management

A user can view, create, modify, and delete Test Requirements.

1. To view a list of Test Requirements, click the List TR's link on the left menu bar. This page lists Test Requirement summaries.
2. To see details of a Test Requirement, click on the link in the ID column. If there are Test Cases associated with the Test Requirement, hyperlinks to them will be shown along with other test requirement details.
3. To create a Test Requirement, click the Register TR link on the left menu bar. The Register Test Requirement page is displayed. The following information should be specified.

   - Organization
   - NDR Version (QOD only)
   - Test Requirement Prefix(QOD only)
   - Test Requirement Number
   - Test Requirement Name
   - Rationale (QOD only) (optional)
   - Test Requirement Status
   - Test Scope (QOD only)
   - Description (optional)
   - Additional Notes (optional)

See the Test Material section for a more detailed description of a Test Requirement.

4. To modify a Test Requirement first follow step 1 and 2 above to find the requirement. Only the owner of the requirement is allowed to modify the

requirement. Hence only the requirement's owner will see modifiable fields and an active update button.

5. To delete a Test Requirement first follow step 1 and 2 above to find the requirement. If the user is the owner of the requirement and there is not a Test Case referencing it, the user may delete it.

3.2   Test Case Management

A user can view, create, modify, delete, and execute Test Cases.

1. To create a new Test Case, click the Add TC link on the left menu bar. The Add New Test Case page is displayed. The list below shows the information that should be specified. Hit the Add Test Case button to submit the new Test Case. Note that if the test case is a script, it will not be executable until it has been approved. When a test is added, an email is sent to the system administrator to approve the script, at which point it will be made executable. (See Test Case section for ways of writing test cases.)

- Test Requirement ID
- Test Case Name
- Description (optional)
- Test Coverage
- Accessibility
- Test Case Type
- Test Script

See the Test Material section for a more detailed description of a Test Case.

2. To view a list of Test Cases, click the List TC's link on the left menu bar. The list of Test Cases page is displayed. The list includes those created by the current user and those that were created by other users for public use.

3. To view the details of a Test Case, click on the Test Case ID hyperlink in the first column of the list of Test Cases. The Test Case Detail page is displayed including the Schematron [5, 8] or Jave Expert System Shell (JESS) [9] script associated with the Test Case.

4. To execute a Test Case, first follow the steps to view a Test Case. There are two ways to submit the data to be tested:
   - copy the target text from a text editor, paste it into the "[XML Schema|XML Document] to be tested" field, and click the "Check" button at the bottom of the page; or
   - select a zip file that contains the XML schema or document to be tested. The zip file may contain one or more files.

   To execute the test select the "Check" button.

The test result is displayed in the web browser and is also stored in the database. The test result can be retrieved by clicking on Previous Test Results on the left menu bar.

5. To modify a test case, first follow the steps to view the test. When the current user is the owner of the test case being viewed, the user can modify the text fields and the "Update" button is available for applying the modifications.
6. To delete a Test Case, from the List of Test Cases page or from the User's Account Page click on the "Delete" hyperlink located on the last column of the table. Only those Test Cases created by the current user have the "Delete" hyperlink. If there is a Test Profile referencing the test case, the test case cannot be deleted (in such cases, the test case owner should use the Test Case status to signify to other users that the test case is deprecated).
7. To execute multiple Test Cases at a time, use a Test Profile. See the Test Profile and the Test Profile Management sections below.

3.3 Test Profile Management

A Test Profile is a collection of Test Cases grouped for a specific set of user requirements. A Test Profile is a mechanism by which a collection of Test Cases are executed at once and the Test Result is stored for subsequent review. A user can create, view, modify, delete, and execute Test Profiles.

1. To view a list of Test Profiles, click the List TP's link on the left menu bar. The User's Test Profiles page is displayed with the list of Test Profiles created by the user, as well as, those created by other users and publicly accessible.
2. To view Test Cases included in a Test Profile, click the ID in the list of Test Profiles. The Test Profile Detail page is displayed showing a list of Test Cases. From here, the user can view Test Case details by clicking on the hyperlinks, as well as, execute the displayed Test Profile (see #6 below for the test profile execution).
3. To create a new Test Profile, click the Create TP link on the left menu bar. The Create New Test Profile page is displayed. A list of Test Cases accessible by the user (user created and public ones) is displayed on the page. Select the check boxes associated with Test Cases to be included in the new Test Profile. Provide a Profile Description at the bottom of the page and click the "Create Test Profile" button. This brings the user back to the User's Test Profile page which now includes the new Test Profile.
4. To modify a Test Profile. Follow steps #1 and #2 above to view the Test Profile Detail. On the list of Test Cases table click the "Remove" hyperlink to exclude the Test Case from the Test Profile. Note that a Test Profile must contain at least one Test Case so the last test case in a Test Profile can not be removed. At the bottom of the list, click the "Insert Test Case into this Test Profile" hyperlink to add Test Cases to the Test Profile. After clicking on the link, a list of accessible Test Cases is displayed. Use the check box on the last column to select Test Cases. Click the

"Insert Selected Test Cases" button to add the selected Test Cases to the Test Profile. This brings the user back to the Test Profile Detail page.

5. To execute a Test Profile, follow steps #1 and #2 above to get to the Test Profile Detail page of the Test Profile you would like to execute. There are two ways to submit the file to be tested:
   - copy the file from a text editor, paste it into the "[XML Schema|XML Document] to be tested" field, and click the "Check" button at the bottom of the page; or
   - select a zip file which contains the file(s) to be tested. The zip file may contain one or more files.

   Click the "Check" button to initiate the execution. After execution finishes (this may take a while), the Test Result page is returned. See the Test Result section for more explanation.

   In QOD only test cases applicable to the selected schema types will be executed against the schema file(s).

6. To delete a Test Profile, follow step #1 above to open the User's Test Profile page. Any Test Profile which the current user is authorized to delete has a "Delete" button located in the last column. Click on the link to remove the Test Profile. After deletion, the user is brought back to the User's Test Profile page. A Test Profile cannot be deleted if there is an associated Test Result. Delete all associated Test Results first, then delete the Test Profile.

The Test Materials section contains more detailed explanation about the Test Profile.

3.4 NDR Profile Management (QOD only)

The functions in this section support the ability to import and export NDR Profiles into an XML file. The XML file is based on the NDR Profile XML schema and is further detailed in a forthcoming user's guide.

Clicking on Export in the menu displays a list of Test Profiles that the user has access to. From this page a user selects a profile and a choice of delivery mechanism (download or email) and an XML file is provided to the user.

When Import Test Profile is selected, the user is shown a screen to guide the import. If the Test Profile is entirely new to the system, meaning that no Test Profile with the same name already exists, and the IDs for Test Requirements in the Test Profile also do not exist in the system, then the import is straight forward. New Test Requirements, Test Cases, and a new Test Profile are all created. However, if the Test Profile or any of the Test Requirements already exist, then the user is presented with choices of how to handle the imported items. The choices are influenced by whether the existing items are owned by the user. The determination as to whether items already exist in the system is based on the IDs used for the items. To facilitate this the QOD interface supports the naming convention described below. If one uses a different interface for creating an NDR Profile instance, it is recommended that IDs are assigned in a similar fashion.

The existence of items in the QOD database is determined by the ID assigned to the item, whether the item is a Test Profile, a Test Requirement, or a Test Case. The Name of the NDR Profile element from the import file is used as the Test Profile ID. When the file is created in QOD, the IDs used for the other items are created as follows. The ID for a Test Requirement is created as the concatenation of the values for

- Organization: the prefix associated with the name of the organization that created the NDR. The values for this field are available from a pull-down menu.

- NDR version: the version number for the NDR version. These values are also available from a pull-down menu.

- TR Prefix (optional): some NDRs provide prefixes for rule identifiers. When these are used in identifiers within the NDR, they will be available from a pull-down menu as well.

- TR Number: when combined with the prefix the TR Number should be unique for the NDR.

Using these values the Test Requirement ID is formed as follows:

guidanceID: organization-NDRversion-[TR prefix]-TR Number

The ID for a Test Case is formed by appending an integer to this ID. For example, in one design guideline document from the Open Application Group (OAGI), *Naming and Design Rules Guidelines Version 9.0* [10], the Test Requirement ID for rule number 16 is OAGI-9.0-R-16. Similarly, the Test Case which implements this Test Requirement is given the ID: OAGI-9.0-R-16-1.

4 Test Materials

This section describes the test materials used by the QOD and the Content Checker tools. Test materials include the Test Documents, Test Requirements, Test Cases, Test Profiles, and Test Results.

Throughout the system the test materials are accessible through listings which contain pointers to pages that detail the specific item. To make these listings more usable a **filter function** is provided via a button at the top of each table. The filter allows the user to view only certain instances of the exhaustive list by applying that filter to the table. The filter page is composed of a set of checkboxes that correspond to the columns in the tables. Different values can be selected or set for the different columns in the table. When the checkboxes are chosen, only those rows which match the column values are displayed in the given table. A note on the page displaying the resulting table indicates what values were filtered for.

4.1 Test Document

The Test Document is either

- an XML schema in the case of QOD or
- an XML document in the case of the Content Checker.

Test Documents are uploaded to the system on either the Test Case or Test Profile Detail pages. See the descriptions of these pages in the Test Procedure section. Both pages allow the user to enter the document onto the screen by typing or using cut and paste. Additionally, the user may upload a zip file containing the document. Note that these are not mutually exclusive ways to provide a document to be tested and both can be used together. In the case of a zip file, only files using the .xsd (for QOD) or .xml (for the Content Checker) extensions are treated; all other files are ignored. If the zip file contains multiple files or if there is an entry in both the text and zip file field, more than one document will be tested. See the Schematron mapping section below to see the implications.

4.2 Test Requirement

Test Requirements in QOD and the Content Checker differ in their origin and nature but, for the purposes of the tools, they have the same characteristics with a couple distinctions as noted below and are handled in the same way. In both tools a Test Requirement is a non-executable statement or assertion. Each Test Requirement may have one or more corresponding Test Cases. However, not all Test Requirements are testable (at least not programmatically testable), so there may be test requirements without Test Cases. A user owns the Test Requirements he/she creates. All Test Requirements are visible to other users, i.e., they are all public.

For the QOD tool many Test Requirements are based on specific best practices drawn from experienced system integrators and/or XML architects. These best practices are often documented in design guidelines (also called Naming and Design Rules - NDRs). The requirements seek to improve the usability, re-usability, and interoperability of the schema by enhancing extensibility, ease of maintenance, implementation and processing efficiency, and the schema's ability to capture and enforce desired semantics.

In the Content Checker, Test Requirements are drawn from the business rules to be applied in specific contexts or transactions. These rules are often not captured in the XML schema being used in the exchange for a variety of reasons. A common reason is that the schema was developed for a more general audience, and the rules only apply in the context of a specific exchange scenario. These requirements supplement the XML schema which supports the definition of the data by enforcing additional constraints on its use.

In both tools the Test Requirement is identified by a unique identifier, refered to as the TRID. In QOD, that identifier is created from the values of the fields for Organization and NDR document version. These values are predefined in the system and assignable through pull-down menus. Alternatively, you can choose "other" as your organization and continue to submit Test Requirements as part of that undetermined organization, but beware that if you chose to associate these Test Requirements with an organization at a different date you will want to recreate them with new ID's. Among other things these ID's will facilitate the Import and Export functions available to support off-line development. In the Content Checker, the identifiers are created by combining the value of the Organization field (assignable through a pull-down menu) with a user-supplied number. If you would like to add organizations or a document version to the system, you will need to contact us at xmlTestbed@nist.gov and we can add your organization and NDR version identifier(s).

In addition to the TRID, the following information is associated with Test Requirements in both tools (unless otherwise noted):

- Organization - This field indicates the origin of the design rule. Test Requirements are usually grouped according to this field. The user needs to contact the tool administrator in order to add an organization to the list.

- NDR Version (QOD only) - This is an identifier for the specific NDR document that is the source for the Test Requirement.

- TR Prefix (QOD only) --- This is an optional field that may be used to qualify the TR number.

- Test Requirement Number (TR Number) - The value of this field should be a number that when combined with the TR prefix is unique within the set of identifiers for a given organization and, in the case of QOD, NDR version.

- Test Requirement Owner (TR Owner) - The value indicates the user who creates and owns the TR. The value is supplied by the system.

- Test Requirement Name (TR Name) - This is a short name for the test requirement.

- Test Requirement Status (TR Status) - The value indicates the life cycle of the test requirement. The status can either be Draft, Active, or Deprecated. The draft status means that the TR is still in a volatile status and suggests that other users may review the TR yet should not reference it. The active status means that the TR is stable, it will not change logically, i.e., only minor editorial changes would be expected. In addition, it implies that the TR is stable enough to be referenced by other users. The deprecated status indicates that the design practice associated with TR is no longer adopted. The owner of the TR uses the deprecated status to indicate such condition to other users who reference the TR through a Test Case (TC). The alternative of deleting the TR would result in the null reference by users of that TR. It should be noted that the system enforces such integrity. That is the TR owner cannot delete the TR if there is one or more TCs referencing the TR.

- Description - The description captures the test requirement assertion (preferably verbatim from a reference document) and provides a detailed explanation about the practice, preconditions, etc.

- Additional Notes - This field should be used to specify information which is not directly derived from the reference document such as how the test can be performed in general, examples of the recommended construct, personal opinions, thoughts, etc.

- Last Update - The Date the TR was last changed. This is a system provided value.

- Derived Test Cases - A list of TCs associated with the TR. A Test Case is assigned to a Test Requirement when the Test Case is created.

- Rationale (QOD only) - This indicates how the requirement supports good design principals. The list below shows example values for this field. These values relate how the rule supports model quality and are applicable in the QOD tool. This is not a restricted list.

    - *Validation and model clarity* is used for those practices that make the semantics of a construct clear to the user as well as to the machine. Subsequently, an XML parser can better validate the content of the instance against the schema.

    - *Structural clarity* is used for those practices that contribute to a schema's readability, which can facilitate consistent interpretation of a schema thereby accelerating its implementation and widespread adoption.

- *Clarity* is used for those practices addressing both the structural clarity and validation and model clarity.

- *Extensibility* is used for those practices promoting reuse through extension. Extensibility also implies reusability.

- *Common symbolic syntax* is used for those practices fostering the use of common naming conventions. Such practices enable better automation and improve readability and clarity.

- *Maintainability* is used for those practices aimed at reducing the maintenance burden especially when changes occur. They help lessen repetitious works and potential errors.

- *Performance* is used for those practices which help reduce the computational overhead associated with the XML instance parsing, validation, and other XML processing.

- *Interoperability* is used for those practices generally promoting the interoperability among partners sharing the same schema. This rationale is very generic and subsumes the first five rationales.

- *Model validity* is used for those practices which help ensure the schema's semantic validity (e.g., no duplication in content). The rules associated with this rationale may overlap with the schema parser or other schema semantic checking tools.

- Test scope (QOD only) - Test scope indicates the types of schemas that the test requirement applies to. One or more types of schemas can be selected. The test scope will help filter out unrelated tests when executing a particular type of XML schema. For example, a weak typing rule, which requires that the primitive schema type should not be used, is applicable only to the aggregate-level and document-level schemas. Specifying that the rule is applicable only to the low-level schema causes the test to be excluded when the schema under test is indicated as aggregate-level or document-level or both. Currently, the QOD tool allows three schema types as described below. (Alternatively Test Profiles can be used to group Test Cases (and hence Test Requirements) for a particular type of schema). More discussion about this is available in the XML Schema Design Quality Test Requirements paper [4].

  - Low-level schema - the schema typically contains simple types and complex types with simple content definitions. This may map to terms in business content standards. For example, the terms can be core-component types, data types, or basic business information entities in the ebXML Core Component specification [11, 12]. They can also be fields, meta, or enumeration in the OAGI specification [13]. These types of schema typically contain reusable, context-free vocabulary, and the elements or types included are not by themselves meaningful to a business exchange.

- Aggregate-level schema - the schema typically contains complex type definitions and corresponding global-element declarations. These may map to terms in business content standards such as the aggregate business information entity in the ebXML Core Component specification or components in the OAGI specification. The constructs in this schema reuse the constructs from the low-level schema.

- Document-level schema - the schema typically contains only a few definitions of complex types and global element declarations. These may map to terms in business content standards such as the assembly document in the ebXML Core Component specification, nouns and business object documents in the OAGI, or the transaction concept in the RosettaNet Implementation Framework specification [14]. The schemas at this level typically do not directly reuse the constructs from the low-level schema but the constructs from the aggregate-level schema.

Test Requirement Example for QOD

Number: 150
Organization: OAGI Design Document
Name: Do not use anonymous type.
Rationale: Extensibility
Test Scope: Aggregate Component, Document/Transaction
Description: Content model of an anonymous type is defined locally within an element. It cannot be referenced outside of that element definition; hence, it cannot be reused. On the other hand, a globally defined type allows it to be referenced and reused. OAGIS adapts the Venetian Blind design approach -- global types should be defined where necessary, but global elements should be declared only for extensible components. See also #300.
Additional Notes: In XML schema, type definitions can be viewed as a content model, but the element definitions are viewed as document structure. The content or data model has a tight relationship with the functional requirements or functional model. Hence, software components should be developed corresponding to the content model rather than to the document structure. The content model represents entities that are used and reused; therefore, software components developed around it can also be reused.

Although the use of global types can cause a name-clashing problem, the availability of namespace mechanisms reduce this problem drastically. The same |

> term with different concepts (perhaps in different domains) can be defined in different namespaces.
>
> Anonymous types may be used for a company's specific terms and terms that have very specific and succinct semantics.

4.3 Test Case

A Test Case (TC) is a computer executable script, i.e., the binding of a Test Requirement to a specific implementation. Generally, a Test Case may contain multiple test steps. Multiple TCs may be associated with the same Test Requirement (TR). Reasons for multiple test cases include different reference data, different scopes, or different rule languages used to implement different aspects of the TR. However, it is recommended that given a choice, a single TC should be used to implement a TR. Limiting the number of test cases simplifies management of the test cases. The user owns the Test Case he/she creates. Only the owner of the TC may delete the TC. If there is a Test Profile referencing a TC, the TC cannot be deleted (see TC status below). A TC may contain the following information.

- Test Case ID - A system-generated unique identification number for the Test Case.
- Test Case Owner (TC Owner) - The value indicates the user who creates and owns the TC. This value is populated by the system.
- Test Case Name - This is a short descriptive name for the Test Case.
- Description - This is an English explanation of the Test Case procedure to help understand the implementation. It may describe test steps, test case execution, other source of information used in the execution, etc.
- Test Script - This field contains the executable rule expression. The executable rule can be expressed either in Schematron or in Java-based Expert System Shell (JESS) language. See the more detail explanation of Schematron and JESS below.
- Associated Test Requirement - This is the Test Requirement ID which the Test Case implements.
- Available for Public Use - This binary field allows the owner of the TC to control the visibility of the TC. If selected, the TC is publicly visible and able to be referenced by other users; otherwise, it is a private TC and is neither visible nor able to be referenced by other users. For example, the user may want to keep the TC private while it is still under development.
- Test Case Type - This indicates the type of rule expression used by the test case. Currently, the tools allow the Test Case to be specified in either Schematron1.5 [15], ISO Schematron [16] or JESS. When the user creates a TC and specifies that it is Schematron, the tool attempts to syntactically validate the script. Validation in the case of JESS is done manually by the system administrator.

Thus for a JESS script there will be a delay from the time it is submitted until it can be executed.

- Test Coverage - The value indicates the extent to which the TC(s) can verify compliance to the test requirement. The possible values are *Full*, *Partial*, or *Unknown*. A TC with a full or partial coverage implies that the TR is *testable*. Full coverage means that the TC can fully verify conformance to the assertion associated with the TR, while the Partial coverage means that the TC can only partially verify the conformance to the assertion associated with the TR. Partial coverage is typically due to technical limitations and/or insufficient information. Unknown coverage means that the test coverage cannot be determined at the time. The reason can be, for example, due to external factors or that the TR itself is ambiguous. It should be noted that the user may create a Test Requirement without an associated Test Case. The reason for this can be that the test requirement is not automatically testable or the definition of the requirement is too ambiguous.

4.3.1 Schematron Considerations

Schematron is a schema language. It can be used to validate rules for any XML document including an XML schema itself. Special considerations for using Schematron in the QOD context include a system to handle multiple schemas during testing and the addition of line breaks to Schematron. These are described below, but first a quick synopsis of the use of Schematron.

The Schematron schema language differs from most other schema languages for XML (e.g., the W3C XML Schema Definition language, Document Type Definition - DTD) in that it is used to specify rule-based assertions rather than grammar expressions. This means that instead of creating a grammar for an XML document, a Schematron schema makes assertions which apply to specific context paths within the target document. If an XML document fails to meet the assertion, a diagnostic message that is supplied by the author of the Schematron is displayed. Schematron is more expressive than the XML Schema language (although more verbose), easy to use, and standards-based. It can validate the content of an XML document and validate relationships between multiple elements/attributes potentially in multiple documents. Its assertion syntax uses the W3C XPath [17] expressions. Further information about the Schematron, including a tutorial, can be found at the following websites:

- The Schematron home page at the Academia Sinica Computing Center [8]
- Resource Directory for Schematron 1.5 [15]
- Schematron Tutorial [18]
- XML.com Introduction to Schematron [19]
- Schematron on Cover Pages [20]
- The Schematron ISO standardization at Schematron.com [16]
- A Schematron schema for RSS [21]

Test Case Example with Schematron Binding
Test Case ID: TC-100
TC Owner: QODMan
TC Name: Improper use of anonymous type
Associated Test Requirement: OAGI-150
Accessibility: Public
Description: This test case verifies that the direct parent element of the simpleType or complexType definition is the schema element. If it is, then it is a global type definition and there is no warning message. If not, a warning message is printed out. Note that the approach that merely checks that the type definition has a name does not work, because a locally defined type may still have a name yet it cannot be referenced (i.e., reused).
Test Coverage: Full
Test Case Type: Schematron 1.5
Test Script:

```
<?xml version="1.0"?>
<schema xmlns="http://www.ascc.net/xml/schematron">
<title>DDMS Test Case 1.1</title>
<ns prefix="xs" uri="http://www.w3.org/2001/XMLSchema"></ns>
<pattern name="OAGI 150: Improper use of anonymous type">
  <rule context="xs:complexType | xs:simpleType">
<!-- The context tells schematron which element(s) to look for. -->
    <assert test="count(../xs:schema)=1">
<!-- This expression checks if the parent element of the context is
the xs:schema element using the count expression. If not, print out
the message below. -->
OAGI-150 [Anonymous Type]: <value-of select="@name"/> element is
locally typed within an element. It cannot be referenced outside of
that element definition; hence, it cannot be reused.
    </assert>
  </rule>
</pattern>
</schema>
```

- **Multiple Schemas**

In order to handle multiple schemas in QOD and the Content Checker, all schemas being checked are concatenated together to make one schema. The system adds an element to enclose the single schema. The user need not do anything to cause this to happen, but the user may need to be aware of the new element when writing rules for these schemas. The root element of the new schema is called QODWrapper and uses UTF-8 encoding[22]. The encoding of all uploaded schemas is checked and only those with the UTF-8 encoding are handled. When multiple schemas are concatenated, the prologues of the individual schemas are deleted to make the final schema well formed.

Example of schema before concatenation

```xml
<?xml version="1.0"?>
<xs:schema xmlns:xs="http://www.w3.org/2001/XMLSchema"
targetNamespace="http://www.w3schools.com"
xmlns="http://www.w3schools.com" elementFormDefault="qualified">
    <xs:element name="person">
        <xs:complexType>
            <xs:sequence>
                <xs:element name="firstname" type="xs:string" />
                <xs:element name="lastname" type="xs:string" />
            </xs:sequence>
        </xs:complexType>
    </xs:element>
</xs:schema>
```

Example of schema after concatenation:

```xml
<?xml version="1.0" encoding="UTF-8" ?>
<qod:QODWrapper xmlns:qod="http://syseng.qod.nist.gov">
<xs:schema xmlns:xs="http://www.w3.org/2001/XMLSchema"
            targetNamespace="http://www.w3schools.com"
xmlns="http://www.w3schools.com"     elementFormDefault="qualified">
    <xs:element name="person">
        <xs:complexType>
            <xs:sequence>
                <xs:element name="firstname" type="xs:string" />
                <xs:element name="lastname" type="xs:string" />
            </xs:sequence>
        </xs:complexType>
    </xs:element>
 </xs:schema>
.....any additional schemas are inserted here....

 </qod:QODWrapper>
```

This allows a user to check some special rules. For example, the user can check that if an assertion is true in one schema or document, it should be true in the other.

The implication for Schematron rules is that, when writing rules using an absolute XPath expression to refer to a node set, the QODWrapper element needs to be referenced and, consequently, the *qod* namespace needs to be declared in the script. In fact, the more specific the path, of which an absolute path is the most specific, the shorter the computation time. Conversely, the use of an asterisk * in the context will result in a long computation time especially when the schema is very big.

Example of schematron script using QODWrapper:

The following example uses an absolute path to check if all elements of the above schema have a name attribute.

```xml
<?xml version="1.0" encoding="UTF-8"?>
<sch:schema xmlns:sch="http://www.ascc.net/xml/schematron">
    <sch:ns uri="http://www.w3.org/2001/XMLSchema" prefix="xs"/>
    <sch:ns uri="http://syseng.qod.nist.gov" prefix="qod"/>
    <sch:pattern name="Check if all elements have a name attribute">
        <sch:rule context="/qod:QODWrapper/xs:schema/xs:element/xs:complexType/xs:sequence/xs:element">
            <sch:report  diagnostics="element-name" test="count(@name)=0" >
                [Error]: The element <sch:name/> doesn't have a name attribute.
            </sch:report>
        </sch:rule>
    </sch:pattern>
<sch:diagnostics>
    <sch:diagnostic id="element-name"> Each element  should have a name attribute
    </sch:diagnostic>
</sch:diagnostics>
</sch:schema>
```

4.3.2 Java-based Expert System Shell (JESS) Considerations

JESS (Java Expert System Shell)[23] is a rule engine and scripting environment written entirely in Sun's JAVA™ language [24] by Ernest Friedman-Hill at the Sandia National Laboratories in Livermore, CA. JESS was originally inspired by the CLIPS [19] (C Language Integrated Production System) expert system shell, but has grown into a complete, distinct, dynamic environment of its own. Using JESS, one can build Java software that has the capability to "reason" using knowledge supplied in the form of declarative rules and facts. JESS is small, light weight, and one of the fastest rule engines available. The JESS engine used is version 7.0, hence JESS script must comply with this version.

Compared with the Schematron, JESS has an advantage of greater expressivity to encode more complex rules, particularly the ability to embed Java calls within the declarative script. The major disadvantage is that JESS does not use the XML object model, in other words, it is not XML native; consequently, it is more difficult to write JESS rules against XML. To use JESS with XML including XML documents and XML schema, the QOD tool transforms the XML using XSLT into JESS facts. The mapping into unordered facts is described below. The person writing JESS rules to check the XML Schema in QOD needs to understand this mapping.

- **Element** -- A fact is instantiated for each element (i.e., each tag) in the XML Schema using this template

(deftemplate element (slot id) (slot name (type STRING)))

The 'id' slot is an arbitrary/auto-generated unique identifier. The 'name' slot is the element tag name. Note that since all element tags in the XML Schema must belong to the W3C

XML Schema namespace (e.g., http://www.w3.org/2001/XMLSchema), the namespace prefix associated with the element name is disregarded.

- **Attribute** -- A fact is instantiated for each attribute in the XML Schema using the following template:

(deftemplate attribute (slot id) (slot name (type STRING)) (slot value (type STRING)) (slot valuePrefix (type STRING)))

The 'id' slot is an arbitrary/auto-generated unique identifier. The 'name' slot is the attribute name. The 'value' slot is the attribute value. It should be noted that all attributes in the XML Schema are considered to have a null namespace; therefore, this fact template does not have a slot for the namespace associated with the attribute name. On the other hand, the value of an attribute may have any namespace association (e.g., an attribute "type" may reference a value "xs:string"). The valuePrefix slot captures the prefix verbatim, if there is one. If the prefix is a default prefix, the valuePrefix slot is assigned "xmlns". If the attribute is not appropriate to have a prefix in its value (e.g., the XML schema attribute "name" cannot have a namespace prefix associated with its value), the valuePrefix slot is assigned "nil".

- **Text** --A fact is instantiated for each text node (text content in between the tags) in the XML document using this template:

(deftemplate text (slot id) (slot value (type STRING)))

The 'id' slot is an arbitrary/auto-generated unique identifier. The 'value' slot is the string content of the text node.

- **Namespace** -- A fact is instantiated for each namespace declaration in the XML Schema using this template:

(deftemplate namespace (slot prefix (type STRING)) (slot uri (type STRING)) (slot parentId (type STRING)))

The 'prefix' slot captures the prefix of the namespace declaration. If it is a default prefix, the value "default" is assigned. The 'uri' slot captures the URI (Uniform Resource Identifiers) part of the namespace declaration. The 'parentId' slot identifies the location of the namespace declaration (which implies the scope of the declaration). The 'parentId' points to the 'id' in an element fact.

- **Relationship** -- A fact indicating containment relationship exists between two elements, an element and an attribute, or an element and a text node is instantiated using this template:

(deftemplate relation (slot type (default attribute)) (slot parent) (slot child))

The 'type' slot can have the value of either 'element, 'attribute', or 'text'. The value of 'element' indicates that the relationship is between two elements. The value of 'attribute' indicates that the relationship is between a parent element and a child attribute. The value of 'text' indicates that the relationship is between a parent element and a child text node. The 'parent' slot is an 'id' from an element fact which is the container element of the relationship, while the 'child' slot is the 'id' from the contained element, attribute, or text fact.

For example, the XML Schema below will generate the following JESS facts.

```
<?xml version="1.0" encoding="UTF-8"?>
<xs:schema xmlns:xs="http://www.w3.org/2001/XMLSchema"
xmlns="http://qod.nist.gov/sample" elementFormDefault="qualified"
attributeFormDefault="unqualified"
targetNamespace="http://qod.nist.gov/sample">
  <xs:complexType name="AmountType">
    <xs:sequence>
      <xs:element name="value" type="xs:decimal"/>
      <xs:element name="unit" type="xs:token"/>
    </xs:sequence>
    <xs:attribute name="qualifier" type="xs:string"/>
  </xs:complexType>
  <xs:element name="Amount" type="AmountType"/>
</xs:schema>

(element (id e1) (name schema))
(attribute (id a1) (name elementFormDefault) (value "qualified"))
(relation (type attribute) (parent e1) (child a1))
(attribute (id a2) (name attributeFormDefault) (value "unqualified"))
(relation (type attribute) (parent e1) (child a2))
(attribute (id a3) (name targetNamespace)
    (value  "http://qod.nist.gov/sample"))
(relation (type attribute) (parent e1) (child a3))
(element (id e2) (name complexType))
(relation (type element) (parent e1) (child e2))
(attribute (id a4) (name name) (value "AmountType"))
(relation (type attribute) (parent e2) (child a4))
(namespace (prefix "xmlns") (uri "http://qod.nist.gov/sample") (parentId
e2))
(namespace (prefix "xml") (uri "http://www.w3.org/XML/1998/namespace")
    (parentId e2))
(namespace (prefix "xs") (uri "http://www.w3.org/2001/XMLSchema")
    (parentId e2))
(element (id e3) (name sequence))
(relation (type element) (parent e2) (child e3))
(element (id e4) (name element))
(relation (type element) (parent e3) (child e4))
(attribute (id a9) (name name) (value "value"))
(relation (type attribute) (parent e4) (child a9))
(attribute (id a10) (name type) (value "decimal") (valuePrefix "xs"))
(relation (type attribute) (parent e4) (child a10))
(element (id e5) (name element))
(relation (type element) (parent e3) (child e5))
(attribute (id a13) (name name) (value "unit"))
(relation (type attribute) (parent e5) (child a13))
(attribute (id a14) (name type) (value "token") (valuePrefix "xs"))
(relation (type attribute) (parent e5) (child a14))
(element (id e6) (name attribute))
(relation (type element) (parent e2) (child e6))
(attribute (id a15) (name name) (value "qualifier"))
(relation (type attribute) (parent e6) (child a15))
(attribute (id a16) (name type) (value "string") (valuePrefix "xs"))
(relation (type attribute) (parent e6) (child a16))
```

```
(element (id e7) (name element))
(relation (type element) (parent e1) (child e7))
(attribute (id a17) (name name) (value "Amount"))
(relation (type attribute) (parent e7) (child a17))
(attribute (id a18) (name type) (value "AmountType")
   (valuePrefix "xmlns"))
(relation (type attribute) (parent e7) (child a18))
```

In addition to the facts produced from the provided XML Schema, the JESS implementation has a "getURI" function defined which provides a short-hand for the user to determine the URI associated with the attribute value. The function takes an id slot value of the attribute fact as an input parameter. The function returns the URI associated with the valuePrefix slot value of the attribute fact. According to the example facts above, calling (getURI a14) would return the XML Schema URI (http://www.w3.org/2001/XMLSchema). If the function cannot resolve the URI for a given attribute fact, it returns FALSE.

A couple more important points about using JESS:

- The output message cannot contain commas, parentheses, and angle brackets (<>). The comma is reserved for formatting the JESS test result. The parenthesis conflicts with the JESS grammar. The angle bracket conflicts with the XML/HTML used for formatting the result.
- Before the user executes the XML Schema against a JESS test case, he/she should ensure that the schema under test is grammatically correct; otherwise, the user risks uncertain XML Schema to JESS translation output which in turn results in wrong test execution.

Test Case Example with JESS script Binding

The example below shows the test case given above in Schematron now written in JESS.

```
;OAGI-150
;Check for Anonymous types
(deffunction isType (?x)
(if (or (str-index complexType ?x) (str-index simpleType ?x))
then
(return TRUE)
else
(return FALSE)
)
)
(defrule oagi-150
(element (id ?e-id) (name ?e-name &: (isType ?e-name)))
(relation (type element) (parent ?p-id) (child ?e-id))
(element (id ?p-id) (name ?e-name2 &: (not (str-index schema ?e-name2))))
=>
(bind ?str (str-cat "" "OAGI-150 [Anonymous Type]: The '"
   ?e-name "' is locally defined within an element. "
   "it cannot be referenced outside of that element definition; "
   "hence it cannot be reused.")
)
```

```
(printout t ?str crlf)
(call ?messagehandler add ?e-id ?str)
)
; Note that the variable ?str should be set to the message provided in the rule.
; The command "(call ?messagehandler add ?e-id ?str)" is used to store each node
; identifier that causes the rule fire with its related message in a class
; designed to handle all these messages.
; With the call of the messagehandler's "add" function, you can give the node ID
; where the rule fired and the relative messages you want to see on the result
; page associated with that node.
; The JESS "printout" function writes the content of this variable on the standard
; JESS output but not on the result page.
```

4.4 Test Profile

A Test Profile is a mechanism for grouping related test cases. For example, one profile might contain test cases for checking a draft schema while another profile groups test cases for checking a schema in its final form. When the user executes a Test Profile, all the included Test Cases are executed. In addition, the Test Results generated from the Test Profile execution are stored in the database for subsequent reviews. The user owns the Test Profile he/she creates. A Test Profile may contain the following information:

- Name - represents a short name of the profile.

- Description - describes its grouping rationale, further information as to who and what should conform to the profile, and/or further instruction to follow after the successful execution of the tests.

- A set of Test Cases - Publicly accessible Test Cases as well as the user's own private Test Cases may be included in the Test Profile.

- Available for Public Use - This field allows the owner of the Profile to control the visibility to other users in the system. If selected, the Profile is publicly available to others for testing their schemas.

Test Profile Example

Profile Name: OAGI Practices
Profile Description: This test profile includes all known test cases associated with the OAGIS design document.
Included Test Cases: OAGI-50, OAGI-100, OAGI-150, ...

4.5 Test Result

Test Results are accessible via the Previous Test Result or My Test Result Page link on the left menu bar. Each of these pages has two tables listing the collection of Test Results stored in the database. The first table contains the results generated from Test Profile execution. The other table contains the Test Result generated from individual Test Case execution. A result generated from the Test Profile execution is stored for 10 days and one generated from a Test Case execution will be stored for 3 days, if the user does not delete them before then. The Test Result is owned by the user who executed the test. Only that user can delete the Test Result. The user can delete a Test Result by clicking on the "Delete" link in the table. These tables contain the following information:

- Test Result: a link to results for a given test execution whether it is a Test Case execution or a Test Profile execution.
    - Test Case execution links directly to the Test Case Results page (see below.)
    - Test Profile Execution links to a table which lists the number of errors for each Test Case. By clicking to the number of errors you will obtain a report which details violations found by the Test Case.
- Tester: the user who executed and owns the Test Result.
- Tested Time: the time when the test was executed.
- Test Profile: a link to the Test Profile used for the test. Links to the individual Test Cases are in the Test Result report.
- A Delete link: Clicking on the link will delete the Test Result from the database. This link only appears if the current user owns the test result.

Test Case Results

The Test Case Results page details the violations for an individual Test Case. At the top of the page is the name of the Test Case which is a hotlink to the Test Case detail page. This is followed by two display boxes. The upper box contains the error listing. When an individual error is selected, the schema segment causing the error is displayed in a lower box.

The error listing box provides the following information for each error:

- The **error number** is on the left.
- The **error message** is created from the test script. Clicking on the error message displays the schema segment where the error occurred.
- The **error location** is the XPpath expression used to find the error in the originating schema file.
- The **schema file name** is the name of the file in which the error was found in the case where the user had uploaded a zip file for testing. When the schema tested was submitted via the screen, this field contains the value "Text Area Document." Sometimes, instead of the schema file name, a link appears with the title *artificial error*. This means that the error points to an element outside the uploaded schema such as the file's name for the schema.

- *(Optional)* The *diagnostic.* This message is optionally implemented in Schematron scripts and provides further detail about the error.

The error listing box contains three errors at a time. Above the box are controls to scroll through the errors.

|<	The first errors
<<	The preceding 3 errors
<	The preceding error
i	The error number i
>	The following error
>>	The following 3 errors
>|	The last error

5 Conclusion

This document describes the operation of the Quality of Design and Content Checker Testing Tools. As of this writing, test profiles are available in the QOD tool based on the following published NDR Documents:

- Department of the Navy XML Naming and Design Rules, Final Version 2.0 [25]

- IRS Enterprise Data Standards and Guidelines: IRS XML Standards and Guidelines [26]

- OAGIS 9.0 - Open Applications Group Integration Specification -- Naming and Design Rules [10]

- UN/CEFACT XML Naming and Design Rules -- Version 2.0 [27]

In the Content Checker test profiles are not based on formal documents but rather the requirements for specific exchange scenarios. Currently there are test profiles available based on an Airforce application for the Department of Defense Metadata Specification [28] and a scenario using the eKanban [29] inventory management system. These tests while less generally applicable are available as examples of how the system might be used in particular scenarios.

6 References

[1] World Wide Web Constortium, World Wide Web Consortium - Web Standards, 2008. Accessed at http://www.w3.org/.

[2] World Wide Web Consortium, W3C Home Page, Accessed at http://www.w3.org/.

[3] World Wide Web Constortium, XML Schema 1.0, 2001. Accessed at http://www.w3.org/XML/Schema.

[4] B. Kulvatunyou and K. Morris, XML Schema Design Quality Test Requirements, NISTIR 7175, October, Accessed at http://www.mel.nist.gov/msidlibrary/doc/Schema_Design.pdf.

[5] B. Kulvatunyou, K. Morris and S. Frechette, Development Life Cycle for Semantically Coherent Data Exchange Specification, Accessed at http://www.mel.nist.gov/msidlibrary/doc/modeldev.pdf.

[6] World Wide Web Constortium, XML Schema, 2001. Accessed at http://www.w3.org/XML/Schema.

[7] MSID XML Testbed Home Page, 2008. Accessed 2008 at http://www.mel.nist.gov/msid/XML_testbed/index.html.

[8] The Schematron Home Page, Academia Sinica Computing Center, Accessed at http://xml.ascc.net/resource/schematron/schematron.html.

[9] E. J. Friedman-Hill, Sandia National Laboratories, Accessed at http://web.njit.edu/all_topics/Prog_Lang_Docs/html/jess/.

[10] G. Minakawa, S. Ramanathan and M. Rowell, OAGIS 9.0 - Open Applications Group Integration Specification--Naming and Design Rules, Accessed at http://oagi.org/downloads/oagis/loadfrm90NDR.htm.

[11] ebXML - Enabling A Global Electronic Market, 2006. Accessed at http://www.ebxml.org/.

[12] United Nations Centre for Trade Facilitation and Electronic Business (UN/CEFACT) Techniques and Methodologies Group, 2005. Accessed at http://www.ifs.univie.ac.at/untmg/.

[13] Open Applications Group, 2008. Accessed at http://oagi.org/.

[14] RosettaNet Home, 2008. Accessed at http://www.rosettanet.org/cms/sites/RosettaNet/.

[15] R. Jelliffe, Resource Directory (RDDL) for Schematron 1.5, 2006. Accessed at http://xml.ascc.net/schematron/.

[16] R. Jelliffe, Schematron, Rick Jelliffe

2008. Accessed at http://www.schematron.com/.

[17] World Wide Web Consortium, XML Path Language (XPath), World Wide Web Consortium, 1999. Accessed at http://www.w3.org/TR/xpath.

[18] M. Nic, Schematron Tutorial, 2000. Accessed at http://www.zvon.org/xxl/SchematronTutorial/General/contents.html.

[19] E. Robertsson, An Introduction to Schematron, Accessed 2008 at http://www.xml.com/pub/a/2003/11/12/schematron.html.

[20] C. Pages, Cover Pages: Schematron: XML Structure Validation Language Using Patterns in Trees, OASIS, 2007. Accessed at http://xml.coverpages.org/schematron.html.

[21] L. Dodds, RSS Validator, Leigh Dodds, 2000. Accessed at http://www.ldodds.com/rss_validator/.

[22] Unicode Inc., UTF-8, UTF-16, UTF-32 & BOM, 2008. Accessed at http://www.unicode.org/faq/utf_bom.html.

[23] E. Friedman-Hill, Sandia National Laboratories, Accessed at http://clipsrules.sourceforge.net/.

[24] Java, Sun, Accessed at http://java.sun.com/.

[25] Chief Information Officer, Department of the Navy XML Naming and Design Rules: Final Version 2.0, United States Department of Navy, 2005. Accessed August 2008 at http://xml.coverpages.org/DON-XML-NDR20050127-33942.pdf.

[26] IRS Enterprise Data Management Office, IRS Enterprise Data Standards and Guidelines: IRS XML Standards and Guidelines, September 15, 2006, Accessed at

[27] United Nations Centre for Trade Facilitation and Electronic Business (UN/CEFACT) Applied Techniques Group (ATG), UN/CEFACT XML Naming and Design Rules -- Version 2.0, 2008, Accessed at http://xml.coverpages.org/UN-CEFACT-XML-Naming-and-Design-Rules-V20.pdf.

[28] Department of Defense Discovery Metadata Specification Home Page, Accessed at http://metadata.dod.mil/mdr/irs/DDMS/.

[29] Electronic Kanban, GS Innovate, Accessed at http://www.gsinnovate.com/PDFs/oneSheets/GSI_eKanban.pdf.

7 Acknowlegements

We would like to acknowledge Lee Ellis at the General Services Administration who provided initial funding for the work, Betty Harvey for her valuable insights and help in defining the needs and export interface, Ken Sall for his encouragement in the early stages of the project, and numerous others who have given their support to this work and without whose help the project would not have been complete.

8 Disclaimer

Mention of commercial products or services in this paper does not imply approval or endorsement by NIST, nor does it imply that such products or services are necessarily the best available for the purpose.

The work described is funded by the United States Government and is not subject to copyright.